Let's Explore Science

EXPLORING THE SOLAR SYSTEM

AMANDA DOERING TOURVILLE

Rourke
Educational Media
rourkeeducationalmedia.com

Scan for Related Titles
and Teacher Resources

www.rourkeeducationalmedia.com

Content Consultant: Diane M. Bollen, Research scientist, Cornell University

Photo credits: Jurgen Ziewe/Shutterstock Images, cover; Noel Powell, Schaumburg/Shutterstock Images, cover; Shutterstock Images, 1, 7, 16, 22, 29; Vasiliki Varvaki/iStockphoto, 4; NASA/AP Images, 5 (top), 8-9 (background), 11, 15, 17; Steven Wynn/iStockphoto, 5 (bottom); Red Line Editorial, Inc., 6, 34; iStockphoto, 8 (foreground), 36, 38; David Gaylor/iStockphoto, 10; TASS/AP Images, 12; Dorling Kindersley, 13, 14 (bottom), 18, 23, 24, 27, 30, 39; AP Images, 14 (top); Lars Lentz/iStockphoto, 19; Mahesh Kumar A/AP Images, 20; Michael Taylor/Shutterstock Images, 21; Shcherbakov Ilya/Shutterstock Images, 25; Martin Adams/iStockphoto, 26; Sabino Parente/ Shutterstock Images, 28, 31, 32; Terry Renna/AP Images, 33; Stephan Messner/iStockphoto, 35; Carolina K. Smith/iStockphoto, 37; Eric Reed/San Bernardino Sun/AP Images, 40-41 (background); Rolf Meier/iStockphoto, 41; Jim McDonald/AP Images, 42; Christian Tatot/AP Images, 43; Scaled Composites/AP Images, 44; Chris O'Meara/AP Images, 45

Editor: Amy Van Zee

Cover and page design: Kazuko Collins

Library of Congress Cataloging-in-Publication Data

Tourville, Amanda Doering, 1980-
Exploring the solar system / Amanda Doering Tourville.
    p. cm. -- (Let's explore science)
Includes bibliographical references and index.
ISBN 978-1-61590-323-8 (hard cover)(alk. paper)
ISBN 978-1-61590-562-1 (soft cover)
1. Solar system--Juvenile literature. I. Title.
QB501.3.T68 2011
523.2--dc22
                              2010009910

Rourke Educational Media
Printed in the United States of America,
North Mankato, Minnesota

rourkeeducationalmedia.com

customerservice@rourkeeducationalmedia.com • PO Box 643328 Vero Beach, Florida 32964

Table of Contents

Chapter 1

SPACE EXPLORATION

People have gazed at the night sky for thousands of years. Ancient peoples used the positions of the stars and planets in the sky to tell the passing of time. Ancient Greeks grouped the stars into **constellations**. In 1610, Galileo Galilei constructed a telescope to see far into the night sky. With his telescope, Galileo discovered more about the universe than anyone had before.

Space exploration has come a long way since Galileo's discoveries. Many spacecraft have been launched into the solar system to photograph and gather information. People have even walked on the Moon.

All of the stars that can be seen in the night sky are part of the Milky Way.

Scientists who go into space are called **astronauts**. Scientists who study space from Earth are called **astronomers**. Astronauts and astronomers work together to gather information about our universe.

Astronomers use many types of math and a science called **physics**. Physics is the study of how the world works. Physicists observe and identify laws of motion and **gravity**. Gravity is what keeps us on Earth. It pulls us down.

Galileo Galilei lived from 1564 to 1642.

Astronomers use the principles of math and physics to learn about the nature of the universe. Because the universe is so large, astronomers have to use what they know about math and physics to create **theories** about what is difficult to observe.

For instance, physicists know that light travels at 186,282 miles (299,792 kilometers) per second. They can use this figure to determine the distance between planets. Using math and physics, astronomers can figure out how long it would take to travel to a distant planet or star using **light-years**. A light-year is the distance light travels in one year.

Let's Calculate How Far Light Travels in a Year

$$\text{speed of light} = \frac{186{,}282 \text{ miles } (299{,}792 \text{ kilometers})}{\text{one second}}$$

Use the speed of light to figure out how far light travels in one year. First, you need to know there are about 31,536,000 seconds in one year.

Next, use the speed of light to figure out how far light travels in one year.

$$\frac{31{,}536{,}000 \text{ seconds}}{\text{one year}} \text{ x } \frac{186{,}282 \text{ miles } (299{,}792 \text{ kilometers})}{\text{one second}} = \frac{5.9 \text{ trillion miles } (9.5 \text{ trillion kilometers})}{\text{one year}}$$

One light-year is equal to about 5.9 trillion miles (9.5 trillion kilometers).

DID YOU KNOW?

Scientists estimate the Sun to be about four and a half billion years old. The Sun is expected to burn much like it does today for another five billion years.

The universe is vast. Huge telescopes can see into other **galaxies** that are billions of light-years away. A galaxy is a large system of dust, gas, and millions to trillions of stars. A solar system is a group of objects, such as planets, that orbit around a star. The Milky Way galaxy is home to our solar system. Earth is part of this system.

Other solar systems exist in the Milky Way galaxy. The Milky Way has hundreds of billions of stars, one of which is the Sun. Any star that can be seen in the night sky is in the Milky Way galaxy.

Our solar system includes the Sun, eight planets, five dwarf planets, and many moons, **comets**, **asteroids**, and **meteoroids**. Planets, dwarf planets, comets, asteroids, and meteoroids orbit the Sun. This means they travel in a path around the Sun. They do so because of the Sun's powerful gravitational pull. A moon orbits a planet because of the planet's gravitational pull. Each planet, dwarf planet, and moon also spins, or rotates, on its own invisible **axis** while orbiting.

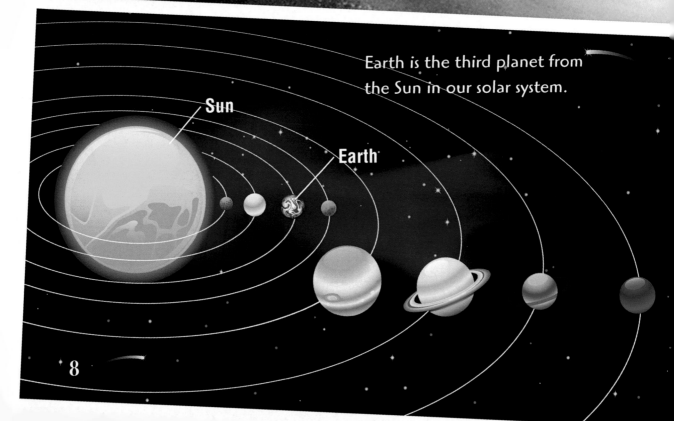

Earth is the third planet from the Sun in our solar system.

Sun

Earth

There are many good reasons to explore our solar system and the space beyond it. Exploring our solar system helps scientists discover our place in the vast universe. By studying other planets, we learn more about Earth and what makes it unique.

Galaxies are grouped by the shape they form. M81 is a spiral galaxy. Other galaxy shapes are elliptical or irregular.

Chapter 2.

WHAT TOOLS DO ASTRONOMERS USE TO STUDY SPACE?

Astronomers need the right tools to do their jobs. They use telescopes, spacecraft, and computers to gather and analyze new information about space.

Astronomers use telescopes to see distant objects in the night sky. Telescopes make objects larger by gathering light from a distant object, focusing the light, and magnifying the image as it enters the eyes.

DID YOU KNOW?

Even the closest galaxies are millions of light-years away. Because it takes millions of years for light from these galaxies to reach us, when we observe these galaxies using high-powered telescopes, we are actually seeing the galaxies as they were millions or billions of years ago.

The first telescopes were very simple and made objects appear about 20 times larger than the original size. This was enough to discover new objects in the night sky.

Today's telescopes have come a long way. They are large and powerful. Ground telescopes are housed in huge buildings. Some telescopes have been launched into Earth's orbit to get closer to far objects. In 1990, astronomers and astronauts put into orbit the most powerful telescope the world has ever seen. The Hubble Space Telescope has discovered new galaxies and produced some of the clearest, most astounding images of space.

The Hubble Space Telescope can see several billions of light-years away.

Sputnik 1

DID YOU KNOW?

In October 1957, the Soviet Union launched
the first artificial **satellite**, Sputnik 1. The satellite
was a sphere with four antennae. Sputnik orbited
Earth in 98 minutes. A month later, the nation
launched Sputnik 2. This time, a dog named Laika
was aboard the satellite.

Many early spectrometers were large and had two arms, like this spectrometer. Some of today's models are very small.

Spectrometers also collect light from objects in space. A spectrometer splits up the light into colors. Astronomers can tell certain details from the light, such as the temperature of an object, which direction it is traveling, how fast it is going, how much it weighs, and what it is made of.

Astronomers gather much of their information from spacecraft. Orbiters that are sent to orbit planets provide information that could not be known otherwise. Some orbiters are manned, but many are not. These orbiters provide astronomers with images of the surfaces of planets and the Moon.

DID YOU KNOW?

In 1958, the United States created **NASA** (National Aeronautics and Space Administration). NASA began working on Project Mercury, the first U.S. human space flight program.

Other useful spacecraft include unmanned probes that are launched into a planet's **atmosphere** or are sent far into space to study the Sun, planets, and other space bodies. Unmanned landers and rovers actually explore the surfaces of planets and moons. They can bring back soil and rock samples for astronomers to study.

Rover

14

More than 15 countries contributed to building the International Space Station.

Astronomers use computers to store and process images and information sent from telescopes, spectrometers, and spacecraft. They also use computers to write software programs to control spacecraft and telescopes.

In 1998, the first piece of the International Space Station was launched. In 2000, the first international crew took up residence at the Space Station. Some crew members stay for several months, doing experiments in space. The Space Station is a step to help scientists learn about the vast universe.

Chapter 3

THE SUN AND TERRESTRIAL PLANETS

Although there are many solar systems, we are most familiar with our own. The Sun is a large star at the center of our solar system. It is a powerful energy source that gives the light and heat needed for life on Earth. It controls the weather in our solar system. The Sun holds the solar system together with its gravitational pull.

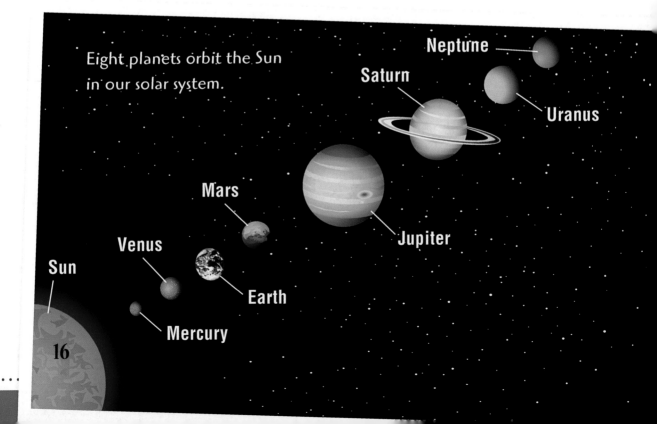

Eight planets orbit the Sun in our solar system.

Neptune

Saturn

Uranus

Mars

Jupiter

Venus

Sun

Earth

Mercury

One million Earths could fit inside the Sun.

Like all stars, the Sun is made up of very hot hydrogen and helium gases. The surface of the Sun is 10,000 degrees Fahrenheit (5,500 degrees Celsius). The core of the Sun is the hottest. It is 27 million degrees Fahrenheit (15 million degrees Celsius).

To us, the Sun is huge, but compared to other stars, the Sun is just average-sized. Other stars are ten times the size of our Sun, but because they are so far away, they appear tiny from Earth. The Sun is about 93 million miles (150 million kilometers) away from Earth. Light from the Sun takes eight minutes to reach Earth. By comparison, light from the next nearest star takes more than four years to reach Earth.

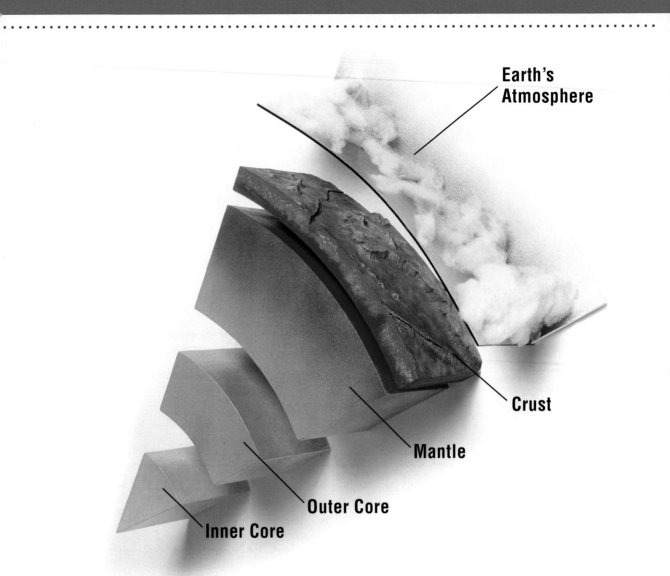

Earth's Atmosphere

Crust

Mantle

Outer Core

Inner Core

Each of the eight planets orbiting the Sun has an atmosphere surrounding it. An atmosphere is a layer of gases. These gases often form clouds, like the ones seen on Earth. The atmosphere helps shield the planet from meteoroids and comets that would collide with it. Most meteoroids burn up in a planet's atmosphere. The atmosphere also helps control the planet's temperature.

The four planets closest to the Sun are Mercury, Venus, Earth, and Mars. These planets have rocky, solid surfaces. They are called **terrestrial** planets, and they are the four smallest planets orbiting the Sun.

MERCURY

Mercury is the first of the terrestrial planets and the smallest planet in our solar system. It is a little larger than Earth's Moon. Mercury has a very thin atmosphere that does not protect its surface. There are many craters from impacts with meteoroids and comets.

Mercury is a very small, dense planet.

DID YOU KNOW?

Planets orbit the Sun and rotate on their axes at different rates. Mercury travels quickly around the Sun in just 88 Earth days, but it rotates very slowly. One Mercurian day is 176 Earth days long.

It takes Venus 225 Earth days to orbit the Sun, but it rotates very slowly. It takes 243 Earth days for Venus to rotate once. Since a day is the time it takes for a planet to rotate once, and a year is the time it takes for a planet to orbit the Sun, a day on Venus is longer than a year on Venus!

Mercury has the widest temperature range of any planet. Because it is so close to the Sun, Mercury's daytime temperatures can reach approximately 800 degrees Fahrenheit (430 degrees Celsius). But because it has very little atmosphere to hold in the heat, Mercury's nights are very cold. Night temperatures dip as low as −280 degrees Fahrenheit (−175 degrees Celsius). Because of the high and low temperatures, life as we know it cannot survive on Mercury.

Besides the Moon, Venus is the brightest object seen from Earth at night.

Venus rotates in the opposite direction from Earth. On Venus, the Sun rises in the west and sets in the east.

VENUS

Venus is the second of the terrestrial planets and the second planet from the Sun. Venus is sometimes called Earth's sister planet. This is because Venus is similar in size to Earth. Venus and Earth are composed of similar elements and have similar gravitational pulls. But the planets have many differences.

A thick, swirling atmosphere surrounds Venus. The pressure of this atmosphere is 90 times the pressure on Earth. This pressure is so high that it would crush metal. The atmosphere is mostly carbon dioxide, a gas that traps heat from the Sun. With temperatures of nearly 900 degrees Fahrenheit (480 degrees Celsius), Venus is even hotter than Mercury. Winds on Venus are about 220 miles per hour (355 kilometers per hour). This is as fast as the winds of a hurricane on Earth.

More than 70 percent of Earth is covered in water.

EARTH

Earth is the third terrestrial planet and the third planet from the Sun. As far as scientists know, it is the only planet with life. It is the only known planet to have liquid water, which living things need.

Humans, animals, and plants are able to live on Earth because of its distance from the Sun. The Sun's heat is enough to keep life warm but not kill it. Earth's atmosphere

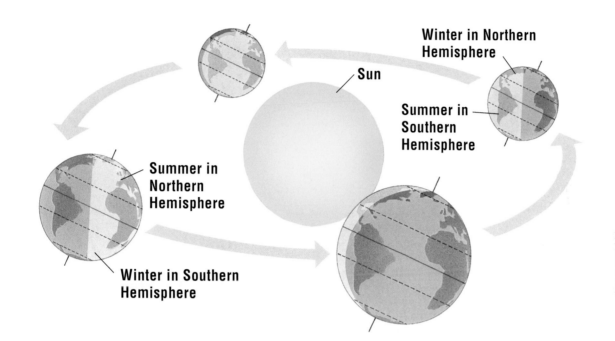

Winter in Northern Hemisphere

Sun

Summer in Southern Hemisphere

Summer in Northern Hemisphere

Winter in Southern Hemisphere

protects life from most of the Sun's harmful rays. The atmosphere traps some of the Sun's warmth so that Earth does not have extreme temperatures like Mercury. But Earth's atmosphere lets out enough of the Sun's warmth so that Earth does not overheat like Venus.

Earth spins on an axis that is tilted at 23 degrees. This tilt gives Earth four seasons. The half of Earth that is tilted toward the Sun experiences more light and the longer days of summer. This time of the year is summer. The other half is tilted away from the Sun and experiences winter.

MARS

Mars is the last of the terrestrial planets and the fourth planet from the Sun. Mars appears red because its soil is rich in iron oxide, or rust. Dust storms are common on Mars. These storms whip the red dust into giant hills. Mars has a very thin atmosphere made of carbon dioxide.

Mars is the planet that is most similar to Earth. Its axis is tilted like Earth's, so Mars has seasons. Mars also has volcanoes, mountains, and canyons. At one time Mars may have had liquid water on its surface. Now Mars is too cold to have liquid water, but ice does exist at its poles.

Scientists are still searching to see if Mars was ever home to tiny life forms.

Chapter 4

THE GAS GIANTS

The four planets beyond Mars's orbit are Jupiter, Saturn, Uranus, and Neptune. These four planets are called gas giants and are larger than the terrestrial planets.

More than 1,300 Earths could fit inside Jupiter.

JUPITER

Jupiter is the fifth planet from the Sun. It is the largest planet in our solar system. Jupiter looks like it is made of light and dark bands of colored clouds. These bands are created by strong winds in Jupiter's atmosphere. Within these bands are storms. One visible storm on Jupiter is the Great Red Spot. Jupiter's core may be solid.

Jupiter's atmosphere is made up of hydrogen and helium. Like Venus, the pressure of Jupiter's atmosphere is enough to crush metal. In 1979 astronomers discovered that Jupiter has three rings around it. These rings are made of small dust particles.

The Great Red Spot has been a raging storm for as long as Jupiter has been observed— about 400 years.

Saturn is the second-largest planet in our solar system.

SATURN

Saturn is the sixth planet from the Sun and is the second gas giant. Because of its beautiful rings, Saturn is called the Jewel of the Solar System.

Saturn's thousands of rings are made up of particles of ice and rock. These particles range in size from that of a granule of sand to that of a house. Astronomers believe these particles are pieces of comets, asteroids, and moons that were torn apart by Saturn's gravity.

Saturn's atmosphere is mostly hydrogen and helium gases. Saturn also has bands of clouds formed by winds, but they are difficult to see. Wind speeds at Saturn's equator rise to 1,100 miles per hour (1,770 kilometers per hour). By comparison, Earth's most devastating tornadoes produce winds of about 300 miles per hour (480 kilometers per hour).

URANUS

Uranus is the seventh planet from the Sun and is the third gas giant. Uranus is very difficult to study because it is so far away from Earth.

Thirteen rings orbit Uranus.

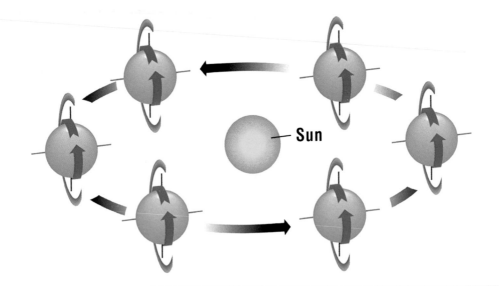

Sun

Uranus's axis is tipped, so the planet rotates on its side. Astronomers believe that a large object may have hit the planet, changing its rotation.

Uranus is made of hydrogen and helium with methane gas in the upper atmosphere. The methane gas gives Uranus its blue-green color. As with Venus and Jupiter, Uranus's atmosphere has high pressure, enough to crush metal. Uranus's atmosphere is very cold, about –350 degrees Fahrenheit (–210 degrees Celsius). The planet's core, however, is very hot. It can reach 12,600 degrees Fahrenheit (7,000 degrees Celsius).

DID YOU KNOW?

A Uranian day is 17.2 hours long, but a Uranian year is very long. It takes 84 Earth years for Uranus to orbit the Sun. Each season lasts more than 20 Earth years on Uranus.

NEPTUNE

Neptune is the farthest planet from the Sun and is the final gas giant. Like Uranus, Neptune is very difficult to study because it is so far away from Earth. Neptune appears blue because clouds of frozen methane gas surround it.

Neptune may be the windiest place in our solar system. Winds on Neptune reach 1,200 miles per hour (1,900 kilometers per hour). Like Jupiter, Neptune has large storms that blow across its atmosphere. Six rings orbit Neptune.

Neptune takes nearly 165 Earth years to orbit the Sun.

Chapter 5

DWARF PLANETS AND MOONS

Until 2006, our solar system had a ninth planet. This small, rocky, icy body was named Pluto. Pluto was the farthest planet from the Sun. It was also the smallest planet.

In 2006, astronomers at the International Astronomical Union (IAU) set criteria to define a planet. Under the new criteria, Pluto no longer qualified as a planet. It was reclassified as a dwarf planet.

Pluto is smaller than Earth's moon.

In 2008, the IAU decided to name all dwarf planets beyond the Neptune orbit **plutoids**. This area is known as the Kuiper Belt. A dwarf planet may or may not be considered a plutoid, depending on where it is located. Five dwarf planets orbit the Sun. Their names are Ceres, Pluto, Haumea, MakeMake, and Eris.

CERES

Ceres is currently the only dwarf planet that is not a plutoid. It orbits the Sun between Mars and Jupiter. Ceres was first discovered in 1801. In 2006, Ceres was classified as a dwarf planet.

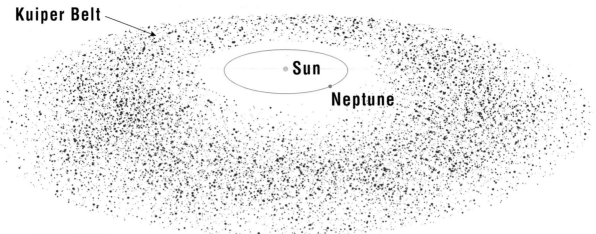

Kuiper Belt

Sun

Neptune

PLUTO

In 2008, Pluto was defined further as a plutoid. Pluto is in an area known as the Kuiper Belt, which is past the orbit of Neptune. Pluto is so far from the Sun that it takes 248 Earth years for it to complete one orbit.

HAUMEA

Haumea was discovered in 2004 and was classified as a dwarf planet in 2008. It orbits the Sun past Neptune in the Kuiper Belt, so it is also a plutoid. Haumea is more oval-shaped than round.

MAKEMAKE

MakeMake was discovered in 2005 and was classified as a dwarf planet in 2008. It is the third-largest dwarf planet and is also a plutoid.

ERIS

Eris was discovered in 2005 and was classified as a dwarf planet in 2006. Because it orbits outside Neptune's orbit in the Kuiper Belt, it is a plutoid. Eris is the largest dwarf planet but is still smaller than Earth's Moon. Eris is the most distant object ever observed to orbit the Sun. It is nearly 10 billion miles (16 billion kilometers) away from the Sun. It takes 560 years to complete one orbit.

MOONS

Moons are natural objects that orbit the dwarf planets and planets. Astronomers have found more than 146 moons that orbit the eight planets in our solar system, but there could be more. Some planets have no moons while others have many.

Earth has one moon. The Moon rotates once on its own axis each time it orbits Earth. Because of this, we see the same side of the Moon all the time.

DID YOU KNOW?

The Moon is not only beautiful to gaze upon, it is helpful too. It helps steady the Earth on its axis, leading to stable climates. The Moon also regulates ocean tides.

Mercury and Venus have no moons. Mars has two moons named Deimos and Phobos. They are very small and are covered with craters.

The gas giants have more moons than the terrestrial planets. Jupiter has more than 60 moons. One of them is the solar system's largest moon, named Ganymede.

The Moon has a very thin atmosphere that offers it little protection from meteoroids. Because of this, the surface of the Moon is covered in craters.

Chapter 6

WHAT ELSE IS IN OUR SOLAR SYSTEM?

In addition to the Sun, planets, dwarf planets, and moons, our solar system is home to millions of smaller bodies that orbit the Sun. These bodies are asteroids, meteoroids, and comets.

ASTEROIDS

Asteroids are chunks of rock made up of the same material as terrestrial planets. These rocky bodies are too small to be considered planets, but they orbit the Sun. Scientists sometimes call asteroids minor planets.

Asteroids range in size from a few feet across (or less than a meter) to nearly 600 miles (965 kilometers) across. While planets are round in shape, asteroids are strange shapes.

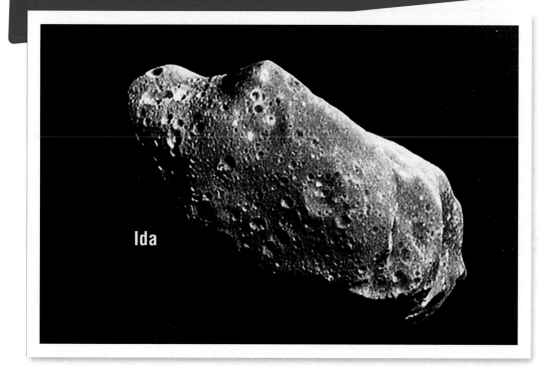

Ida

Most of the solar system's asteroids can be found in the asteroid belt orbiting between Mars and Jupiter. However, stray asteroids have been found. Asteroids sometimes hit planets. Scientists monitor asteroids that come near Earth in case one gets too close.

METEOROIDS, METEORS, AND METEORITES

Meteoroids are smaller than asteroids. Meteoroids are made up of rock and metals. When a meteoroid enters Earth's atmosphere, it is called a **meteor**. Meteors can start out quite large, but Earth's atmosphere causes them to burn up. This is why they appear to glow. Most meteors burn up before reaching Earth, but sometimes they fall to the surface. When they do, they are called **meteorites**. Rarely, these meteorites cause impact craters in Earth's surface. There is one large crater in the state of Arizona on the continent of North America. It is about three-fourths of a mile (1.2 kilometers) across.

Meteor showers are seen as streaks through the sky.

Meteors can be random, but some meteor activity can occur during the same time each year. Meteor showers are times when meteor activity increases. These showers appear to come from the same spot in the sky. They are named by the star constellation from which they seem to appear. The most famous meteor shower, Perseids, can be seen every year around August 12. It is most visible in the Northern Hemisphere.

Meteorite

The comet Hyakutake was seen over Frankfurt, Germany, in 1996.

COMETS

Comets are bodies of rock and ice that orbit the Sun. Their orbits are more oval-shaped than the orbits of planets, dwarf planets, and asteroids. Comets consist of a solid core and a hazy atmosphere. As comets approach the Sun, they heat up, trailing gas and dust behind them. The Sun illuminates this tail, making it glow. One popular comet is Halley's Comet. It can be seen in the night sky every 76 years. The last time it appeared was in 1985 to 1986. The next chance to view the comet will be in 2061.

FUTURE POSSIBILITIES

Astronomers and astronauts have learned much about the universe and they will continue to study space. NASA has a goal of returning people to the Moon by 2020. There are also hopes that the Moon can be used as a staging point for future human flights to Mars.

DID YOU KNOW?

Most comets are not bright enough to be seen without the aid of binoculars or telescopes. Comets that can be seen with the naked eye are called great comets. One of the last great comets to be seen from Earth was Hale–Bopp in 1997. Hale–Bopp moved slowly across the sky and was visible for 15 months. Because Hale–Bopp is a long-period comet, it will not be seen again for more than 2,300 years.

Hale–Bopp

In the future, a vacation may be a trip into space. In 2001, a man paid 20 million dollars to be taken to the International Space Station. Several companies are now working on spacecraft to take private citizens into space. These flights will be very expensive, about $200,000 in U.S. currency. Companies also hope to open a hotel in space.

In 2004, SpaceShipOne became the first private spacecraft to reach space.

NASA's space shuttle program launched its first mission in 1981. Even if the program ends, it does not mean an end to space exploration or the operation of the Space Station. Other countries will continue providing astronauts with transportation to the International Space Station.

Glossary

asteroids (ASS-tuh-roidz): large chunks of rock that orbit the Sun; scientists believe asteroids are material left over from the formation of the solar system

astronauts (ASS-truh-nawtz): people who go into space to study Earth or other bodies in space

astronomers (uh-STRON-uh-merz): scientists who study space from Earth

atmosphere (AT-muhss-fihr): a mass of gases that surrounds a planet or moon

axis (AK-siss): an invisible straight line around which a planet or moon rotates

comets (KOM-itz): bodies of rock and ice that orbit the Sun; when comets are near the Sun, they trail dust and gases that are lit up by the Sun, giving them the appearance of a tail

constellations (kon-stuh-LAY-shuhnz): groupings of stars

galaxies (GAL-uhk-seez): very large groups of stars and associated bodies of space

gravity (GRAV-uh-tee): the force that pulls people, planets, and moons in a certain direction; the Sun, Earth, and other planets have gravitational pull

light-years (LITE-yihrs): the distance light can travel in one year; light travels at 186,282 miles per second

meteor (MEE-tee-ur): a meteoroid that enters Earth's atmosphere

meteorites (MEE-tee-ur-ritez): meteors that reach Earth's surface without burning up

meteoroids (MEE-tee-ur-oydz): smaller chunks of rock and metal that orbit the Sun; meteoroids are similar to asteroids but smaller

physics (FIZ-iks): the study of how the universe works; how matter and energy interact

plutoids (PLOO-toydz): dwarf planets that are farther than the orbit of Neptune

satellite (SAT-uh-lite): a body orbiting another; satellites can be natural, like the Moon, or artificial, like spacecraft

terrestrial (tuh-RESS-tree-uhl): relating to Earth; in space, planets that are like Earth

theories (THIHR-eez): ideas that are based on fact and observation but have not been proven

Index

Websites to Visit

Kids Astronomy. www.kidsastronomy.com

NASA Kids' Club. www.nasa.gov/audience/forkids/kidsclub/flash/index.html

NASA Science for Kids. http://nasascience.nasa.gov/kids

Solar System Exploration. http://solarsystem.nasa.gov/kids/index.cfm

About the Author

Amanda Doering Tourville is the author of more than 50 books for children. She hopes that children will learn to love reading as much as she does. When she's not writing, Amanda enjoys reading, traveling, and hiking. She lives in Minnesota with her husband.

Meet The Author!
www.meetREMauthors.com